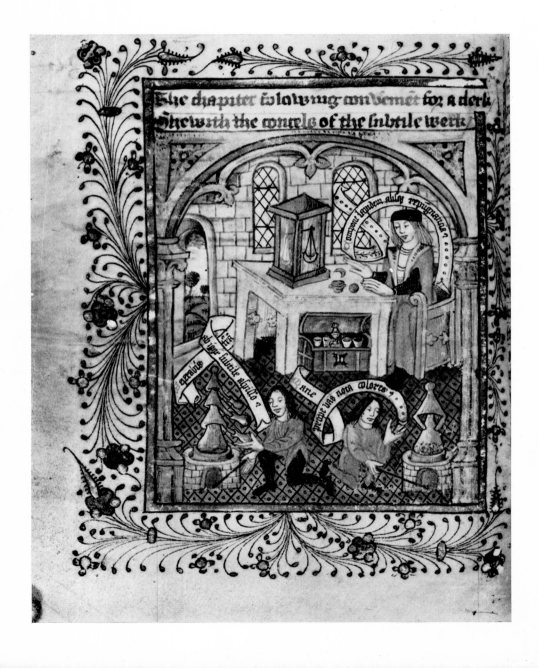

1 A fifteenth-century alchemist at work
(Reproduced by permission of the British Museum)

A Science Museum Survey

Development of the Chemical Balance

by **John T. Stock**
Ph.D., D.Sc., F.R.I.C.,
Professor of Chemistry
University of Connecticut,
Storrs, Connecticut, U.S.A.

London
Her Majesty's Stationery Office 1969

Foreword

This short history of the chemical balance, by Professor John Stock of the University of Connecticut, is based on a study of the Chemical Balance Collection at the Science Museum. With few exceptions, the instruments referred to are on permanent exhibition. Those which are not are available for inspection at reasonable notice. We are grateful for those colleagues of other museums who have allowed their instruments to be included for the sake of completeness.

Professor Stock's essay is part of a series being arranged by Mr. Frank Greenaway, Keeper of the Department of Chemistry, in which, as opportunity allows, visiting scholars will be invited to examine critically collections of those instruments which have played a notable part in the development of chemistry, and are particularly well-represented at the Science Museum. It is hoped, for example, that the collections of hydrometers and of heating devices will eventually be the subjects of monographs similar to this one.

The Science Museum is indebted to Professor Stock both for the light he has thrown on many familiar instruments and for the suggestions he has made for further study.

It is also a pleasure to thank a valued friend of the Museum, Mr. J. R. Waite of L. Oertling Ltd., for his advice throughout the progress of this work.

D. H. Follett
Director

Contents

1 The Importance of Weighing in Chemistry 2

2 Factors Governing the Sensitivity of a Balance 6

3 Early Balances 9

4 The Rider System 26

5 Weights and Buoyancy Effect 28

6 Modern Balances 30

7 Microbalances 36

8 Recording Balances 44

References 48

Appendix 50

1 The Importance of Weighing in Chemistry

It is not known when the first weighing was carried out. The ancestor of the steelyard, a weighing device with only one pan, seems to have appeared first. Although the two-pan equal-arm scale, or balance (Latin, *bi-lanx,* meaning 'two dishes'), may have developed later, it has a history of more than three thousand years and was certainly known to the ancient Egyptians. Balances are depicted in various alchemical writings; the earliest representation of a balance in a case is probably in an illustration of a 15th-century alchemical laboratory (Figure 1)[1]. Balances capable of weighing small loads with reasonable precision were in existence several hundred years ago, and were used by early jewellers, money changers, and metallurgists.

Although the early metallurgical assayers habitually performed weighings, it was Joseph Black, Lecturer in Chemistry at Glasgow and then at Edinburgh University, who first provided examples of reasoned sets of chemical experiments in which the balance was used at almost every stage[2]. In the period 1752–55, Black proved that limestone (calcium carbonate) changed into quicklime (calcium oxide) when 'fixed air' (carbon dioxide) was removed, and that limestone could be obtained from quicklime by a reversal of this process. He drew similar conclusions about the constitution of magnesia alba (magnesium carbonate) and some other carbonates. These discoveries, significant in themselves, were doubly important because they were quantitative. Black's results, given to the nearest grain (one grain = 64.8 milligrams = 0.00229 ounce), were probably obtained with a fairly crude balance. The Science Museum has a replica (Figure 2) of a balance known to have been used by Black; the original is in the Royal Scottish Museum, Edinburgh. It is just possible that this balance is the one actually used by Black in his work on limestone. The great French chemist Antoine Laurent Lavoisier (1743–1794) was the first to stress the paramount importance of quantitative studies and the consequent need for really sensitive balances. From the outset of his career, Lavoisier fully believed in the principle of conservation of mass. This means that although chemicals may react to form new substances, the total weight remains unchanged. It is interesting that Lavoisier's own results justify this fundamental chemical principle to no better than about one per cent[3]. Although the much more accurate results of many subsequent workers do not conflict with this principle, the experiments were not generally designed to test the principle itself; the validity of the principle was assumed to be unquestionable. Direct proof was provided by Hans Heinrich Landolt (1831–1910), whose study of the problem extended over the last twenty years of his life. Using a balance that could weigh a load of 500 grams with a maximum error of a few thousandths of a milligram, Landolt could detect no certain change in weight caused by any of the reactions that he studied[4]. The principle of conservation of mass, proved experimentally to hold to within one part in a million, is therefore presumed to be exact. Although this principle applies to

TESSERACT

Box 151
Hastings-on-Hudson
New York 10706
(914) 478-2594

Early Scientific Instruments

David Coffeen, Ph.D.
Yola Coffeen, Ph.D.

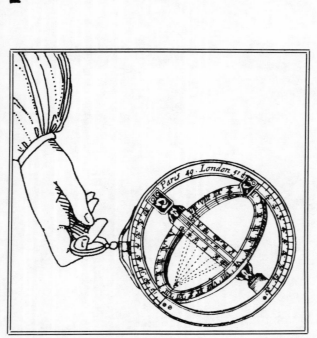

30 Nov 82

Dear Mr. Kennedy,

Thank you for your order of the 27th.

Please do let me know if you think of anyone else who might like my catalog.

We are also active buyers of good early instruments, if you should come across such.

Sincerely,

Gary Coffin

2 Copy of a balance used by Joseph Black

3 Balance used by John Dalton

the almost limitless number of purely chemical reactions, it does not necessarily hold when the nucleus of an atom undergoes a change. In nuclear reactions, which are often associated with the release or uptake of enormous amounts of energy, it is the total of mass and energy that is conserved.

During the first decade of the 19th century, a notable controversy occurred between the French chemists Claude Louis Berthollet (1748–1822) and Louis Joseph Proust (1755–1826). Berthollet maintained that when elements unite to form a compound, their proportions may vary within quite wide limits. Proust assiduously collected evidence to show that a particular compound always contains the same elements united together in the same proportions, a statement later known as the Law of Constant Composition. He also noted that when two elements, such as oxygen and iron, form more than one compound, then the composition of each compound is fixed. That such a controversy should have occurred is indicative of the state of analytical chemistry at the time. Data that appeared to support Berthollet's view could readily be found, while experimental errors of at least one per cent occurred in Proust's work. Although the Law of Constant Composition is a fundamental principle in most aspects of modern chemistry, it is interesting that the existence of certain variable-composition solid compounds is now well authenticated. Such compounds are aptly termed berthollides[5]. (It has been appreciated since the formulation of the Law of Mass Action that Proust and Berthollet were really arguing at cross purposes).

During the period of this controversy, John Dalton (1766–1844) enunciated the Atomic Theory. According to this, an element is composed of identical atoms that differ from the atoms of any other element. Compounds are formed by the union of two or more different atoms. Dalton stressed the importance of knowing the relative weights of the various atoms and gave a list of these in which the hydrogen atom was assigned unit weight. He assumed that if two elements, A and B, form only one compound, then this will be AB. The only nitrogen-hydrogen compound known to Dalton was ammonia, so he assigned it a composition (in modern terminology) NH. Dalton's value of 5 for the atomic weight of nitrogen is therefore roughly one-third of the modern value (corresponding to NH_3). Many smaller discrepancies occur; although Dalton made outstanding contributions to chemical theory, neither was he a gifted experimentalist nor had he refined equipment at his disposal. His balance, shown in Figure 3, was a simple construction and was not encased. This instrument passed into the possession of the Manchester Literary and Philosophical Society, of which Dalton was president for many years. It was destroyed by aerial bombing during the Second World War. A pocket balance (Figure 4) in the Science Museum closely resembles a similar instrument owned by Dalton.

The formulation of the underlying quantitative principles of chemistry showed up the deficiencies of existing analytical procedures. With great patience and consummate skill, the Swedish chemist Jöns Jacob Berzelius (1779–1848) developed improved methods of purification and analysis, determined the atomic weights of some forty elements, and ascertained the composition of many compounds. He demonstrated the convenience and accuracy of analytical procedures involving the weighing of quite small quantities. It is from his time that the chemical balance began to depart from the massive form of many of its ancestors and to take on its modern appearance and capacity. The successors of Berzelius, notably Jean Baptiste Dumas (1800–1884), Jean-Servais Stas (1813–1891), Theodore William

Richards (1868–1928), and Otto Hönigschmidt (1878–1945) reduced the uncertainty in the value of many atomic weights to the order of one part in many thousands. Such results could not have been attained without the development and use of highly-accurate weighing devices and techniques.

Today, mass spectrometry has largely replaced quantitative chemical analysis as a means for the determination of atomic weights[6],[7]. The need for accurate and rapid chemical analyses is however greater than ever, so that the analytical balance is still a major and fundamental tool of the chemist.

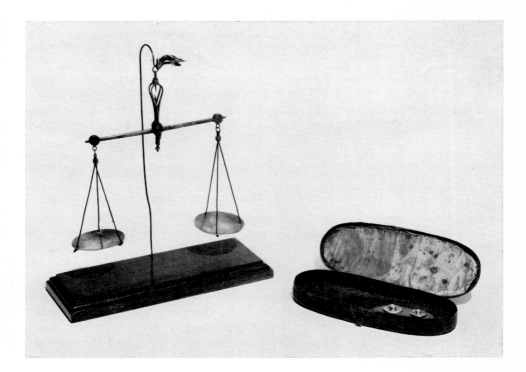

4 Pocket balance and case, early 19th century

2 Factors Governing the Sensitivity of a Balance

5 Beam of a precision balance, showing knives

Although precision balances differ greatly in size and in design, their performances are controlled by a common set of factors. The mathematical theory of the balance is well established[8, 9]; a simplified treatment that concerns only the sensitivity is sufficient for present purposes.

Ideally, the edges of A, B and C (Figure 6) of the three knives should be parallel and in exactly the same plane, while the half-beam lengths AB and BC should be identical. An important part of the balance maker's art is the close fulfilment of these conditions. Mounted centrally on the beam a little in front of centre knife B, the long pointer is perpendicular to the plane ABC. A small scale set behind the fine tip of the pointer enables the deflection or oscillation of the beam to be observed. Since the maximum deflection of the pointer from its vertical position is never allowed to become large, the scale is only about an inch in total length.

The modern balance beam shown in Figure 5 has a somewhat unusual mounting of the 'pointer' that allows all three pivots, which in this case are knives of agate, to be clearly shown. The form of beam combines rigidity with comparative lightness. The pointer is replaced by a graticule (see Chapter 6).

The deflection is greatly exaggerated in Figure 6, which illustrates in a simplified fashion the underlying principle of the balance. Force f, due to the combined masses of the load on the left-hand pan, the pan itself, and its suspension system, tends to turn the beam in a counterclockwise direction. This tendency is opposed by the two forces f' and w, the first of which refers to the right-hand pan, its accessories, and the weights or other objects on this pan. If the pans and their accessories are identical, (f − f') represents the difference in the loads on the pans that causes the beam system to be deflected through angle a. Force w is due to the mass of the beam system, including the pointer, and may be considered as operating through point G, the centre of gravity of the beam system. This point must lie *below* the edge B of the central knife, or the balance will be unstable and the beam will flop to one side or the other instead of swinging. Although real, the distance BG in a sensitive balance is very small.

The opposing turning effects depend both upon the forces and the points of application of these forces. A given force applied some distance from the fulcrum B obviously has a greater turning effect than if this force is applied very near the fulcrum. The relationship is quantitative if the distances from the fulcrum are measured horizontally, since all of the forces operate vertically. If the beam system is to come to rest in the position shown, the following relationship must be obeyed:

$$f \times FB = f' \times BD + w \times BE$$

Since the half-beam lengths are equal and, by definition, the ratio FB/AB = cosine a, and BE/BG = sine a, the relationship becomes

$$\frac{(f - f') \times AB}{w \times BG} = \frac{\text{sine } a}{\text{cosine } a} = \text{tangent } a$$

Because the maximum deflection permitted in a modern balance is small, tangent a is closely proportional to a convenient practical quantity, namely, the deflection expressed in scale divisions.

AB, BG, and w are all fixed in a balance of ideal construction, so that the deflection, in scale divisions, is closely proportional to the difference between the loads on the pans (the pans, with accessories, are assumed to be identical). A modern pointer-type analytical balance will accommodate an object weighing not more than about 200 grams, yet an out-of-balance (f − f'), of only one milligram will cause a deflection of several scale divisions. For example, a particular balance may have a sensitivity of 4.0 divisions per milligram or, as sometimes expressed, 0.25 milligram per scale division. Since readings can be estimated to a fraction of a scale division, an out-of-balance of one tenth of a milligram can easily be detected.

In these considerations, the beam is supposed to have been allowed to come to rest. Deflections are often observed with the beam in motion, several successive scale readings being taken as the pointer changes its direction of travel[10]. This procedure avoids the waiting time needed for motion to cease and, since the small frictional forces at the knife edges are minimised when the beam is in motion, gives more reproducible results. However, certain modern balances incorporate damping devices that cause the beam to move to its equilibrium position with almost no oscillation.

An obvious way of obtaining high sensitivity (i.e., a large deflection for a small difference in the loads on the pans) is to use a long, light beam. Even on this basis, a compromise will be necessary, because the longer the beam, the heavier it will be. Furthermore, a balance having a very long beam will swing very slowly, especially if adjusted to give high sensitivity. Such an instrument will be tedious to use, especially when many weighings have to be made. Lightness can be improved by careful design and by the use of light alloys, but no material is absolutely rigid. The beam therefore bends more and more as the loads on the pans are increased. This causes the distance BG (Figure 6) to become greater, so that the sensitivity is not constant, but diminishes as the loading is increased. The effect can be very

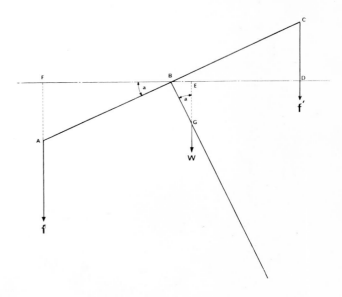

6 Representation of the forces acting upon the beam of a balance

pronounced if the balance has a long, thin, rod-like beam and is adjusted to be highly sensitive in the unloaded condition.

Certain early balances, such as those of the Ramsden pattern, have beams that are designed for all-round rigidity and are consequently quite heavy. Since the forces that cause bending operate downwards and not sideways, it is obvious that rigidity in the vertical plane is particularly important. Some early instruments, and all modern ones except certain special devices for very small loads, have beams that are quite thin back to front, but of considerable width top to bottom. All unnecessary metal is removed and a triangulated design is often employed, so that the beam is as light as is compatible with a high degree of rigidity. Nevertheless, the change of sensitivity with loading is more noticeable with a long-beamed balance than with an instrument having a shorter beam of similar design. Modern practice is to use a quite short very rigid beam and to obtain high sensitivity by making the distance BG very small.

An obvious way of eliminating these changes in sensitivity is to operate under conditions of constant loading. This principle, discussed on page 34, is employed in the design of modern high-speed balances.

7 Egyptian cord pivot balance, 14th century BC

3 Early Balances

The pans of very early balances were often suspended from cords that passed through holes in the ends of the beam (Figure 7), or were merely tied to the ends. Passage of time brought improvements, an important one being the introduction of the beam with swan-neck ends (described in 1574 by Ercker). Although the date of this development is uncertain, balances with this type of beam were in use by the 16th century (Figure 8). An incomplete 'Hydrostatick Balance,' belonging to St. John's College and on loan to the Museum of the History of Science, Oxford, has a beam with swan-neck ends and is one of the oldest existing balances used for scientific purposes. The instrument, which is incomplete, is attributed to Francis Hauksbee, who died about 1713, but is neither signed nor dated. It is similar to the 'Hydrostaticall-Ballance' described by Hauksbee in 1710[11].

The swan-neck design allowed the easy approximate equalisation of the effective half-beam length by bending the extremities towards or away from the centre of the beam, as indicated at (a) in Figure 9. In a modified form, apparently due to

8 Reconstruction of a 16th-century assayer's balance

Kater[12], the bending is closely controlled by screws that are mounted at right angles. This arrangement provides for both the equalisation of the half-beams and the lining-up of the end bearings with the central one. The pan supporting cords are usually attached to a metal ring or hook that rocks on the bottom of the slot in the swan neck. Although absent in the balance used by Black (Figure 2), better instruments of the swan-neck type have a knife edge formed at the bottom of the slot, as shown enlarged at (b) in Figure 9. The fulcrum or central bearing of the beam is usually a steel knife that passes transversely through the beam. The edge of the knife rocks in holes near the ends of a stirrup-shaped support that straddles the beam. A string attached to this support enables the balance to be held by hand or otherwise suspended when a weighing is to be performed (Figure 4). A major advance in balance design was the introduction of the knife and plane system, which is shown schematically as an end bearing at (c) in Figure 9. When a centre bearing is of this type, the edge of the knife is of course downwards and the plane is beneath the knife. The simple geometric form of each of the two pieces involved allows their finishing to a high degree of perfection, while the line of contact between knife and plane can be of considerable length. For a given load, the greater force per unit area of contact in the hook and swan-neck design, or in any other essentially point-contact type of pivotal system, naturally implies a greater rate of wear than will occur in a knife and plane system.

Some early balances have knives and planes that are both of hardened steel. Other balances use steel knives in conjunction with agate planes. The first suggestion to use agate planes as bearings in a balance appears to have been made by Henry Cavendish (1731–1810)[13]. Steel knives may of course rust, and they can give

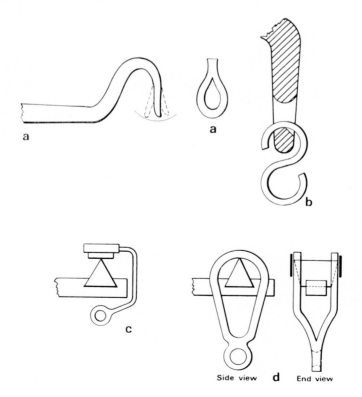

9 Typical balance end-bearing systems
 (a) swan-neck end; the adjustment, by bending, of the effective half-beam length is indicated by dotted lines
 (b) section through slot of swan neck
 (c) knife and plane bearing
 (d) side view and end view of loop-type bearing

rise to uncertainties in weighing due to magnetic effects. The first use of agate as a material for both knives and planes has been attributed to Robinson[14], whose activities are described on page 22.

Precision balances made after about 1770 usually have a knife and plane centre bearing. However, end bearings of the loop or roof type were popular until at least half a century later. The loop-type bearing, shown schematically in side view and end view at (d) in Figure 9, actually has two loops, one on either side of the beam, that are joined together at the eyelet. It is often difficult to see the actual bearing surfaces, because the loops are partially closed by steel retaining plates. (In Figure 9 (d), these plates are omitted from the side view, so that the actual bearing can be shown). These plates restrict the front-to-back movement of the loops on the knife edge. To minimise rubbing against the retaining plates, the ends of the knife are chamfered as indicated in the end view.

Since wear can take place only when the components of a bearing are in contact, a precision balance is provided with an *arrestment* which enables the components of the bearings to be kept apart except when an observation is being made. This mechanism is usually operated by a knob or lever on the outside of the case, so that the actual weighing can be done under draught-free conditions with the case closed. Many early balances have a single-action arrestment that merely separates the components of the centre bearing. In some cases there is a separate control to lift the pans, so that the residual loads upon the bearings at the ends of the beam are quite small when the balance is out of use. Some later instruments employ a triple-action arrestment that is operated by a single control. As this is rotated, the central bearing, then the end bearings, and finally the pans, are arrested. This sequence is reversed when the control is turned the other way to release the moving parts.

The shape of a roof-type bearing and the associated retaining plates give this bearing its self-locating property. Since a knife can remain in contact with any portion of a plane, a knife and plane bearing has no power of self-location. The release of the arrestment must therefore perform the additional function of causing each knife to take up a fixed position upon its plane. Otherwise, the pointer attached to the beam will not stand truly on the pointer-scale, or might actually rub against this scale; the knives would work towards the extremities of the planes and then be damaged in the resulting spillover.

The early balance shown in Figure 10 is a copy of the original (now in the Royal Institution) made to the instructions of Henry Cavendish[15, 16, 17]. Particularly notable is the use of the knife and plane system for both centre and end bearings. This balance was constructed by Harrison — presumably John Harrison (1693–1776), so famous in connection with the development of the marine chronometer. Although its date is uncertain, this balance may be the one used by Cavendish in his work on the composition of water, carried out between 1781 and 1784[18]. From his account it appears that small loads could be weighed to about one-tenth of a grain, or about six milligrams. An examination of the original balance in 1965 showed that the performance was still at least as good as this.

With an overall height of more than five feet, the case and stand of the balance are of unprepossessing appearance and belie the excellence of design of the mechanism within. Constructed from sheet iron, the rectangular-section beam is 19.5 inches long. Small brass blocks and screws allow vertical adjustment of the steel centre knife, which turns on steel planes. Since the steel end knives are also provided with screw adjustment, the half-beam lengths can be closely matched. A small weight mounted on a screw projecting downwards from the underside of the beam allows the sensitivity of the balance to be altered at will. Each pan is suspended from a brass universal joint (Figure 11). This is attached to a curved steel bracket the upper part of which is formed into a pair of planes. Two V-shaped crutches arrest the beam by lifting it at both ends. These crutches, which also ensure that the end knives lie over the centre lines of their respective planes (Figure 12), are attached to vertical brass tubes that are caused to slide up and down by cams operated through gearing from outside the case. When required, additional support for the pans is given by brass pins that are moved by two levers.

Lavoisier owned precision balances made by Pierre Bernard Mégnie (c. 1751–1807) and by Nicolas Fortin (1750–1831). Lavoisier believed that there were no balances other than those made by Ramsden that could match the accuracy and precision of his Fortin instruments[19].

Jesse Ramsden (1735–1800) (Figure 13) was born in Yorkshire and came to London as a cloth warehouse clerk in 1755[20]. He entered the workshop of an instrument maker named Burton some three years later, then set up his own business about 1762. His inventive genius and great skill soon gained him an international reputation; he was elected a member of the Royal Society in 1786 and of the Imperial Academy of St. Petersburg in 1794. The demand for Ramsden's magnificent

10 Copy of a balance owned by Henry Cavendish

astronomical and other instruments was greater than could be satisfied by the output of sixty workmen. He invariably rejected any completed instrument that fell short of his ideal, and often appeared to be unconscious of the passage of time. Once on being commanded to appear before George III, the king had occasion to remark that Ramsden was punctual as to day and hour, but late by a whole year! Despite his great activity, Ramsden's disregard of gain led to but a small fortune. The fine Ramsden balance shown in Figure 14 was used in experiments made in 1789 to determine accurately the specific gravities of water-alcohol mixtures[21]. This work was the result of an appeal by the Government to Sir Joseph Banks, then President of the Royal Society, for the best method of proportioning the excise on spirituous liquors. The pans of the balance are now missing, but it could carry a considerable load and then turn to about one-hundredth of a grain, or about half a milligram. Two hollow brass cones that are joined at their bases form the beam, which carries steel knives. The central knife turns on agate planes that are fixed on a heavy brass frame supported by four brass pillars, while hangers with loop-type bearings are provided for the end knives. Two V-shaped supports, operated through a screw and lever system by a small handle outside the case, lift the beam by making contact near the centre, as indicated in Figure 15. Although this arrestment system is more compact than the arrangement used in the Cavendish balance, the lifting of the beam near the centre has at least one disadvantage. Slight wear or distortion of the lifting gear can give rise to a corresponding uncertainty

11 Pan suspension device of Cavendish's balance, showing universal joint and planes for end knife

in placing the central knife on its planes. Because of their much greater distance from the lifting gear, the lateral uncertainty in the location of the pointers at the ends of the beam is much greater. When the Ramsden balance was examined in 1965, the left hand pointer often rubbed against its scale when the beam was released.

Fine balances of design similar to that of Ramsden were made by other instrument makers such as Robert Fidler and Edward Troughton, as well as in the Holborn workshop of the brothers William and Samuel Jones. Balances by all of these makers are in the Science Museum collections. The Ramsden type of balance often has a small screw on the upper centre of the two-cone beam. Rotation of this screw causes the rise or fall of a weight that is contained within the beam, and thus allows the sensitivity of the balance to be adjusted.

A balance made by Fidler for the Royal Society, and now in the Science Museum, carries a light brass finely-divided scale on the right-hand end of the beam. A hole is cut in the glass of the case in front of this scale. Although these modifications may have been made after the instrument had left Fidler's workshop, they appear to constitute one of the earliest examples of provision for the reading of a balance with the aid of a telescope or low-power microscope.

Edward Troughton (Figure 16) was born in Cumberland in 1753[22, 23]. In 1770 he became an apprentice mechanician to his brother, John, who had a business in London. When John died a few years later, Edward Troughton carried on the business

12 Action of the arrestment of Cavendish's balance

13 Jesse Ramsden
(Reproduced by permission of the Royal Society)

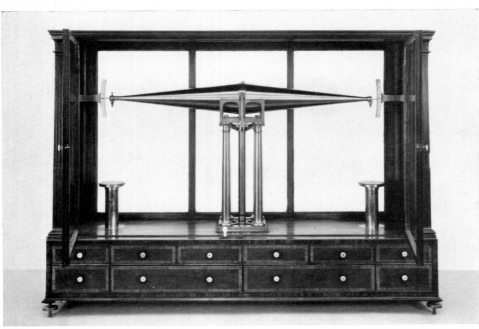

14 Balance constructed by Ramsden

16 Edward Troughton
(Reproduced by permission of the Astronomer Royal)

15 Beam lifting arrangement of Ramsden's balance

at 136 Fleet Street first alone, and then from 1826, in partnership with William Simms. Like Ramsden, Troughton led a simple and frugal life. He was less concerned with profit than with his reputation, which was truly international. His surveying and other instruments travelled to many parts of the world, and were used in the American coast survey of 1815. Troughton was elected to the Royal Societies of London and of Edinburgh in 1810 and 1822 respectively, and he died in 1835. His name was for long perpetuated in the firm of Cooke, Troughton, and Simms, Ltd., now incorporated in Vickers Instruments, Ltd., of York.

In 1798 Sir George Shuckburgh Evelyn published an account of his endeavours to set up standards of weight and measure. His account includes a description of the Ramsden-type balance with a beam 27 inches long that was made by Troughton for this work[24].

Another Troughton balance, made for Sir James South, is shown in Figure 17. The mahogany beam is over two feet long and turns on conical steel points. An arrestment, in the form of two pins that enter hollows in the underside of the beam, is actuated by a spring-loaded brass rod. Depression of the spring allows the conical points to make contact with small circular agate plates. A platform beneath each pan is attached to a spring-loaded brass tube that can move up and down in a vertical tubular guide. Closure of the hinged lid of the balance case depresses the two brass tubes and hence frees the pans. Each pan hangs from thin cords that pass over a brass knife edge at the end of the beam (Figure 18).

When South married an heiress in 1816, he gave up his extensive practice of surgery to devote himself to astronomical and other scientific activities[25]. In the observatory that he fitted up at his home in Blackman Street in South London, South had a transit made by Troughton. South eventually acquired a twelve-inch object glass, then the largest in existence, and instructed Troughton and his partner Simms to

17 Balance made by Troughton for Sir James South

build this into a telescope. When the work was completed in 1831, South complained that the mounting lacked rigidity and refused to pay. Since he remained extremely bitter after a judgement in favour of Troughton and Simms, he is unlikely to have ordered further equipment from these makers. It thus appears that the balance was constructed within the period covered by the two dates mentioned. The balance shown in Figure 19 is of German origin and is inscribed 'F. W. Breithaupt & Sohn in Cassel.' It was used by Dr. Andrew Ure (1778–1857), whose 'Dictionary of Arts, Manufactures, and Mines' was once well known. Frederick Wilhelm Breithaupt (1780–1855) was the son of the instrument maker Johann Christian Breithaupt, who set up his business in 1760[26]. The triangulated beam of this balance is built up from strips of brass and carries three steel knives. Agate planes are mounted on a block, which is caused to rise by rotation of the knob at the front of the instrument. The planes thus meet the central knife and then lift it, so that the beam is well above its arrested position and clear of the stops that project from either side of the centre column. Figure 20 shows one of the hangers that support the pans. The single bridge-shaped retaining plate has been removed in order to expose the roof-type bearings. In balances of the Ramsden type,

18 Pan suspension arrangement of the balance made by Troughton

19 Balance used by Andrew Ure

20 Pan hanger, with retaining plate removed to show roof-type end bearing

21 Open-beam balance with swan-neck end bearings
(Reproduced by permission of the Museum of the History of Science, Oxford)

22 Robinson balance used by Wollaston
(Reproduced by permission of the Whipple Science Museum, Cambridge)

lateral movement on the knife is restricted by separate small retaining plates on the outer sides of each hanger (Figure 9 (d)).

An inventory of 1823 lists several balances contained in the Ashmolean Museum's Chemistry Laboratory[27], which is now the basement of the Museum of the History of Science, Oxford. Little further is known of the origins of these balances, one of which is shown in Figure 21. This balance incorporates a peculiar combination of design features. The open-form flat brass beam combines lightness with resistance to bending under load and carries a triangular-section steel centre knife. An unusually slender centre column carries quartz or glass planes and, near its bottom, the arrestment control lever. When this lever is released, the beam is lifted near its ends by vertical projections from a horizontal bar, so that the knife no longer touches the planes. A V-notched slide then centres the knife above the planes. These fairly advanced features are offset by the use of swan-neck end-bearings, which are of steel and are shaped to accommodate the extremities of the beam.

Thomas Charles Robinson (1792–1841), of Devonshire Street, Portland Place, London, was the first manufacturer of precision balances[12, 28]. Instruments constructed by him were used in the adjustment of copies of the Imperial Standard troy pound in 1825[29] and again in 1829[30]. The standard was destroyed in the fire that swept the Houses of Parliament in 1834 and had to be re-established. It is clear from an account of this work[31] that although Robinson had passed away by then, his balances and his designs had not. Apparently, the first precision balances used in the United States were those of Robinson[28].

Existing Robinson balances have a surprisingly modern look, with a short beam of perforated triangular design. Inclined slits near each end of the upper side of the beam are provided with screws that permit the final equalisation of the lengths of the arms and the lining-up of the knives. The pointer is quite short and projects

23 Robinson balance used by Miller
(Reproduced by permission of the Whipple Science Museum, Cambridge)

24 Ludwig Oertling
(L. Oertling, Ltd.)

downwards from the centre of the beam to move over a scale that is mounted about halfway up the central column. A small weight travelling on a screwed portion of the pointer stem permits the sensitivity of the balance to be changed. Robinson adjusted his $10\frac{1}{2}$-inch beam balances to detect a weight difference of one-thousandth of a grain with a load of one thousand grains in each pan, or one-millionth part of the weight to be determined[32]. These particular balances have steel knives that work on agate planes. The beam and end planes are located by a crutch system similar to that used in the Cavendish balance, but the operating mechanism of the arrestment is much simpler than in the Cavendish instrument. The pans are supported by silk threads and can be arrested by moving separate levers.

Two Robinson balances are in the Whipple Science Museum of the University of Cambridge. Although very small — the beam length is only $5\frac{1}{2}$ inches — they are generally similar to Robinson's larger instruments. The first (Figure 22) has associations with and was used around 1820 by William Hyde Wollaston (1766–1828), who discovered and isolated the precious metals palladium and rhodium. This instrument has steel knives and an agate centre plane. However, the end planes are of steel and are cut away in the centre so that they bear only on the ends of their knives. The other balance, which belonged to William Hallowes Miller (1801–1880), Professor of Mineralogy at Cambridge[31], is shown without its case in Figure 23. Its steel knives make full-length contact with agate planes. Each end plane is attached to the underside of a small rectangular brass plate. A single short j-shaped wire terminating in a ring to take the pan threads projects downwards from the plate.

Ludwig Oertling (1818–1893), whose portrait appears as Figure 24, was the founder of the present-day firm of L. Oertling, Ltd. He was born in Berlin and, before going to London, was apprenticed to his brother. Since the firm has records going back to 1847, Oertling's British balance-making activities cannot have started later than this. For some time after Robinson's death, Oertling was almost the only balance maker to retain the system of knives and planes throughout, and not to use roof-type bearings for the end knives[33].

Figure 25 shows a balance that was used in 1896 by William Ramsay and Morris W. Travers in their work on neon and the other rare gases of the atmosphere. This type of balance, first manufactured by Oertling about 1847, continued in production until some years after the end of the First World War. 'Oertling long beam' balances achieved an international reputation for precision and durability; for example, the Russian chemist Dmitrii Ivanovich Mendeleev, of 'Periodic Table' fame, used one of them[34]. Despite their age, quite a number of these balances are in good working order today. One of the pan hangers of such a balance is shown in Figure 26; the agate plane is clearly seen. This arrangement is similar to the one used by Robinson.

25 Late 19th-century balance used by Ramsay and Travers

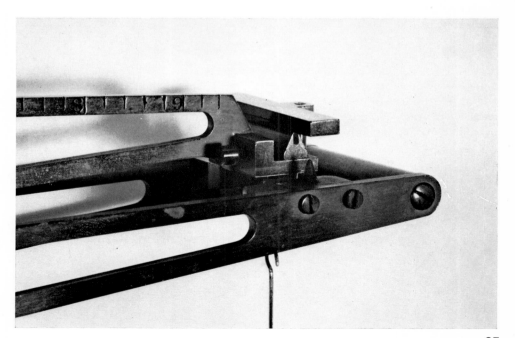

26 Pan hanger of Oertling 'long beam' balance, showing agate plane

4 The Rider System

Although a chemical balance will turn to a small fraction of a milligram, weights as small as this are impractical to handle. Instead, a single weight of more convenient size is used, and is moved along the beam as needed. This special weight, which is usually of wire and in the form shown in Figure 27, is appropriately known as a *rider*. Commonly, the right half-beam is divided into ten major equal parts (often both half-beams are thus divided), with further subdivisions into a total of fifty or one hundred parts. Sometimes the beam carries a special rider track on which the markings appear. The main divisions are numbered from the beam centre, so that '10' falls over or alongside the right-hand knife. It is obvious that a 10-milligram rider placed at the reading main divisions, as shown in Figure 28, will have the same turning effect as weights to a total of milligrams that are placed on the right-hand pan.

27 Ten milligram rider, shown approximately full size

Since the zero of the rider scale is over or alongside the centre knife, the balance can be adjusted for even swinging before the rider is put in place. If the balance has a very short beam, the zero and the '10' of the rider scale are often found at opposite ends of the beam. A 5-milligram rider is then used, and the balance is adjusted to swing evenly with this rider in place at the zero mark. Since the effect is to oppose the weights on the right-hand pan when the rider is on the left half-beam, and to augment them when the rider is moved to the other side of the centre knife, the 5-milligram rider used in this manner can provide an effective range of ten milligrams. With a rider of any size, provision is made for adjusting its position from outside the case. As in the changing of weights, movements of the rider are made only after the balance has been arrested.

It is difficult, if not impossible, to assign credit for the introduction of the rider system. However, on the occasion of the Great Exhibition, held in London in 1851, and at which most of the leading makers of balances were exhibitors, L. Oertling received a Council Medal (the highest award) for his balance 'with graduated beam and sliding apparatus'[35].

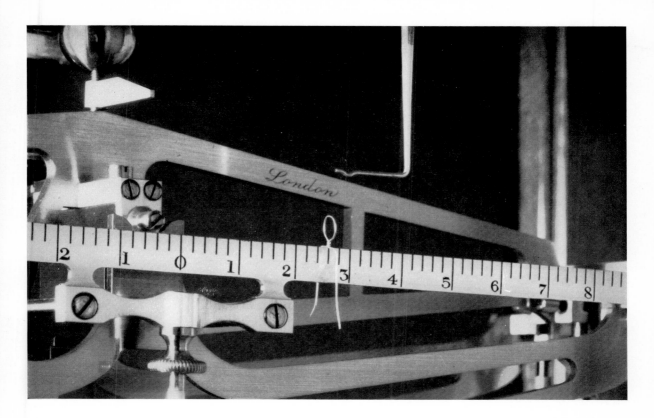

28 Rider in place on rider scale of balance

5 Weights and Buoyancy Effect

No balance can produce results better than permitted by the weights used with it. Figure 29 shows a set of grain weights used around 1824 by W. H. Wollaston. The inscription on the lid of the box shows that the weights had been carefully calibrated against Troughton's standard weights.

Nowadays, the metric unit of weight, the gram, is used universally except for certain highly-specialised purposes such as the weighing of precious stones. Although acquired by the Science Museum as long ago as 1876, the set of weights shown in Figure 30 is generally similar to sets that are manufactured at the present time. In the set shown, the weights from one gram upwards are of gilded brass. Nowadays, harder alloys are often used. The fractional weights are made from platinum (sometimes aluminium) sheet. All weights are handled only with forceps having ivory or similar tips. To facilitate such handling, a knob surmounts each cylindrical weight, while each fractional weight has a bent-up edge or corner. Unless there is some special reason for departing from the convention, the weights are placed on the right-hand pan, and the object to be weighed is kept on the left-hand pan.

Since object and weights are rarely of the same density, they displace different volumes of air and are therefore subject to different buoyancy effects. The apparent weight of the object will therefore differ a little from the true value. The error is normally small enough to be ignored in routine analytical work and can usually be allowed for by calculation when the requirements are more stringent.

Although rarely used except for weighings of the highest accuracy, operation in vacuum is an obvious way of completely eliminating the buoyancy correction, and also such effects as differential adsorption of moisture on the weights and on the object to be weighed. Manipulation is, however, more difficult, and the case of the balance must be both vacuum-tight and capable of withstanding considerable external pressure. A vacuum balance, made by Oertling to the design of W. H. Miller and now in the Science Museum's Metrology Collection, has $1\frac{3}{4}$-inch thick glass plates at the front and back of the case. This instrument was used from 1872 to about 1892 for accurately comparing copies of weights with the primary standards. The pans and their loads can be interchanged without destroying the vacuum. so that allowance can be made for any slight difference in the half-beam lengths of the balance.

29 Sets of grain weights used by Wollaston

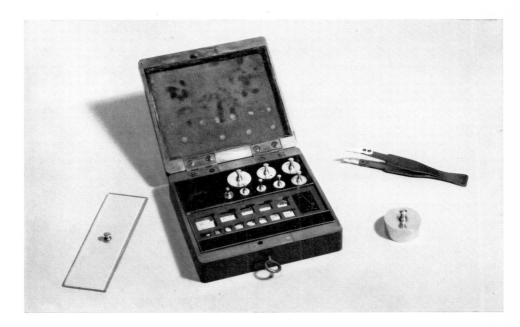

30 Set of 19th-century metric weights

6 Modern Balances

The balance has a long and continuous history, so that any attempts at classification into early and modern forms must be on a purely arbitrary basis. The one adopted here is to assign the origin of the modern balance to the year 1866, when Paul Bunge (1839–1888), a German engineer who was interested in the design of bridges, introduced the short-beam analytical balance. His aim was to increase the speed of weighing by reducing the time required for a complete swing of the pointer.

Short-beam balances existed long before Bunge's time, but were mostly of the non-precision type (as, for example, the pocket balance shown in Figure 4). A notable exception is the $5\frac{1}{2}$-inch beam Robinson balance used by W. H. Miller, (Figure 23), which is definitely a precision instrument[31]. It was Bunge, however, who developed the theory and practice of short-beam instruments[36].

Bunge became attracted to the subject of balance design through discussions with an instrument maker friend, who held the then prevailing view that high sensitivity was best achieved by the use of a long beam. Disputing this view, Bunge showed clearly that any sensitivity lost by shortening the beam can be amply regained by improvements such as in the finishing and mounting of the knives. High sensitivity without loss of stability can then be obtained by raising the centre of gravity of the beam and its accessories until almost coincident with the edge of the centre knife. After serving an apprenticeship under Frederic Apel, instrument maker at the University of Göttingen, Florenz Sartorius (1846–1925) in 1870 founded the business that still bears his name. He was the first to manufacture a short-beam balance with a beam of aluminium, having access to this then-new metal through his countryman, the famous chemist Wöhler. The Sartorius balance (Figure 31) acquired by the Science Museum in 1876 has a $5\frac{1}{2}$-inch long triangulated aluminium beam, steel knives, and agate planes. When examined in 1965, this balance was found to allow a weight difference of one-tenth of a milligram to be detected when the load on each pan was one hundred grams. The balance will therefore discriminate to one part in a million. Such a performance is routinely expected of any balance that is used for chemical analysis. More modern balances usually have agate knives that bear on agate or, since about 1950, on corundum (synthetic sapphire) planes.

Many improvements in balance design have been directed towards ease, and hence speed, of weighing, rather than further increasing sensitivity or accuracy[37, 38]. The 'chainomatic' principle, used earlier in other connections but patented with respect to analytical balances by Christian A. Becker (1874–1946) in 1915, is a good example of this, Neither rider nor any weight smaller than a decigram is used, the equivalent effect being obtained by means of a fine chain. One end of this is suitably hung from the right arm of the beam, while the other end is attached to a carriage that travels vertically on a graduated column, as shown in Figure 32.

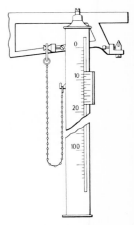

32 Arrangement of tractional-weight chain and column (E. Mettler Ltd)

31 Nineteenth-century Sartorius balance

This carriage, which has a vernier to permit readings to one-tenth of a milligram, is operated from outside the balance case. Since the turning force on the beam increases as the carriage is lowered, the column is numbered downwards. In other designs, the chain is wound on or off a drum that carries a suitable scale.

Some of the sensitive balances included in the Great Exhibition of 1851 carried a small mirror on the beam, so that the reflected image of a fixed scale could be viewed through a telescope. A beam of light then becomes equivalent to a very long pointer of zero weight and infinite rigidity; this principle is a feature of many modern balances. One method of application in balances used for standards work is to illuminate a fine slit, shown diagrammatically as an arrow in Figure 33, so that

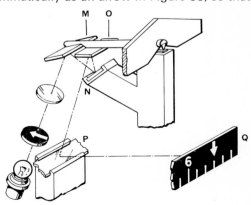

33 One form of optical projection system

the lights falls upon a mirror M carried on the beam. After being reflected by this mirror, the fixed mirror N, a second moving mirror O, and a second fixed mirror P, the light forms an image of the slit on the rear of a translucent scale Q. A more usual method is to project the greatly-enlarged image of a tiny scale on a ground-glass screen that is located for easy viewing by the operator. The popular method is to project an image of the scale itself, the scale being attached to the pointer or beam. A 2 mm. graticule may be magnified 40 to 80 times.

The speed with which routine weighings can be carried out with a simple precision balance is limited by the time required for the addition or removal of weights and for the positioning of the rider or other final-adjustment device. Many modern balances have built-in partial or complete sets of weights that can be applied to or lifted from the beam system by controls on the outside of the balance case. This design possesses advantages beyond the mere quickening of weighing. Being inside the case, the weights cannot be dropped, worn by careless handling, or accidentally mixed with those from another set. The controls usually take the form of knob-operated dials on which the weight readings appear directly in bold figures; no mental arithmetic, however simple, is required. In the Sartorius balance shown in Figure 34, the principle is applied to fractional weights only, so that external weights from one gram upwards are also required. A later balance (Figure 35) by the same maker requires no external weights. This instrument has only one pan, which is used to carry the object to be weighed. In place of a second pan, there is a set of horizontal bars to carry the mechanically-applied weights.

The most obvious (and, incidentally, the most accurate) method of using a balance

34 Sartorius balance, with damping device and internal fractional weights

35 Sartorius 'Selecta' balance

is to operate it as a null-point device. The aim is to counterbalance exactly the object to be weighed, so that the pointer oscillates about its rest position. A balance designed to operate as a deflection device permits weighings to be performed more rapidly. With such an instrument, the final decimal places of the result are read directly from the extent of the displacement of the beam or pointer, instead of from the position of the rider or similar device. Optical projection is nearly always used so that reading is simple.

In the design and construction of a deflection-type balance, some important requirements must be met. The sensitivity must remain constant at all loads within the capacity of the balance, and must be adjusted so that the scale reads directly in mass units. For example, a deflection of one scale division might indicate an out-of-balance of exactly one milligram, so that the weighing could be estimated to the nearest one-tenth of a milligram. Further, damping must be incorporated to cause the natural oscillations of the beam and its appendages to die out almost instantaneously, so that a steady scale reading is obtained within a few seconds of the release of the beam. This type of instrument is sometimes termed an aperiodic balance. Eddy-current magnetic damping is sometimes used, but pneumatic damping is much more common. A light piston A, of large diameter, is attached to one end of the beam, as shown in Figure 36. The piston fits cylinder B fairly closely, but does not touch it. This cylinder is shown in section in the figure, and is mounted on the wall of the case or on some other fixed part of the balance structure. Since air can escape or enter only through the annular space between A and B, the device opposes any rapid motion of the beam. However, this retarding effect decreases to zero as the beam comes to rest. Some balances have damping devices at both ends of the beam.

Although the actual location on the beam varies with the design, thumbnuts C and D are encountered in practically all kinds of modern balances. The rotation of C causes the centre of gravity of the beam system (and hence the sensitivity of the balance) to be raised or lowered. This adjustment is normally made only by a balance expert. Thumbnut D allows the position of equilibrium, or 'rest point,' to be adjusted, so that the pointer swings approximately equally on either side of the centre line of the scale when the pans are empty. The left arm of the beam usually carries another thumbnut matching that shown at D.

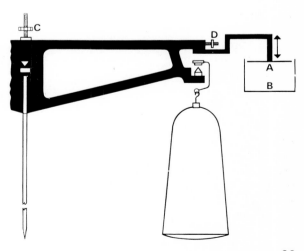

36 Arrangement for damping the oscillations of a balance

The problem of maintaining a sensitivity that is completely independent of the mass of the object to be weighed is best solved by operation under constant load. The principle is best understood by considering the little-used 'substitution method' of operating a simple balance. All of the weights from the accompanying box are placed on the left pan and the weights are counter-balanced by a piece of metal (or even a small vessel containing lead shot) that rests on the right pan. The object to be weighed is then placed upon the left pan, thus upsetting the counterbalancing. Weights are now progressively removed from this pan until the state of equilibrium is restored, when the result is found by counting up the masses of the discarded weights. Whatever the mass of the object, the balance carries a constant load. Since they are not touched after the initial counterbalancing, the right-hand pan and its load can be replaced by a block of metal of the same total mass, so that a 'one pan' balance is obtained. In fact, the extra metal can be incorporated in one arm of the beam, so that only two knives, A and B, are required (as first popularized by Mettler of Zurich). The weights C need not actually rest upon the one remaining pan, but can be placed upon a subsidiary platform that is attached to the pan support, as shown diagrammatically in Figure 37. This platform usually takes the form of a series of horizontal bars, upon which ring-shaped weights are hung. Knob-operated dials that control a system of cams enable various combinations of weights to be lifted from or replaced on the bars. The result is indicated directly by the dial settings, with aperiodic readout of the final decimal places. Figure 38 shows the arrangement of weights and lifting mechanism of an Oertling constant-load balance. The essential features of such a balance are more clearly indicated in Figure 39.

The simplicity of operation, speed, and precision of a constant-load balance are achieved by careful design and construction. Because the load is kept at the maximum for which the balance is designed, excessive wear of the knives and planes is a distinct possibility in a constant-load balance of poor design. Wear is minimised by increasing the length of the knife that is in contact with the plane. Ideally, weights should neither be lifted not be reapplied unless the beam is arrested; otherwise, damage to the knives and planes may occur. However, rapid weighing is impossible unless the selection of weights to be lifted can be made quickly. A two-stage arrestment is sometimes used partially to reconcile these conflicting needs. Initially, the beam is freed, but its movement is restricted to within very small limits that are sufficient to permit the selection of weights. The arrestment control is then moved to its second position, in which the beam is freed completely and can move to its equilibrium position. A more fundamental approach, employed in Oertling constant-load balances, is the incorporation of a preweighing device. When this is operated, the entire instrument acts as a spring balance with a full-range optical readout. This gives the mass of the object with an accuracy sufficient for the setting of the weight dials. This setting is done after reapplication of the arrestment; an interlocking device prevents the inadvertent changing of any of the larger weights while the beam is free. On its second release, the beam moves to its final position for readout of the complete result.

Although sometimes outwardly similar, the balances of the various makers differ considerably in constructional details. Technical information is given by specialist monographs[38, 39] and, to some extent, by trade literature.

38 Internal weights and weight-operating mechanism of a constant-load balance

39 Exposed view of the working parts of a constant-load balance

37 Essential features of a constant-load balance

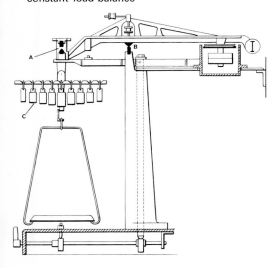

7 Microbalances

A balance that will weigh an object of up to about 200 grams to the nearest one-tenth of a milligram is suitable for most chemical laboratory operations that require precise weighing. An exception occurs when very small amounts of materials are involved. A *microbalance* is then employed. One form is essentially a miniature analytical balance of maximum capacity from 10 to 20 grams, but with a sensitivity so high that a difference of only one microgram (one-millionth of a gram) can be detected. The need for a comparatively large capacity arises from the fact that chemicals are never placed directly upon the balance pan, but are weighed in suitable containers. A good example occurs in the determination of carbon and hydrogen in an organic substance such as paraffin wax[40]. A sample of a few milligrams is weighed into a little platinum boat and is burned in a stream of oxygen. The gas stream, which now contains the carbon as carbon dioxide and the hydrogen as water vapour, is passed in succession through two previously weighed absorption tubes. The first of these retains the water and the second the carbon dioxide. When the combustion is complete, the tubes are reweighed. Although each tube may weigh 10 grams or more, the results are calculated from the very small *increase* in weight due to the compounds retained. For this reason, a set of microchemical weights contains small fractions of a gram only. Most of the counterbalancing is done with a counterpoise such as a spare absorption tube.

Some balances made before the Second World War had beams only 3 inches long. Although sensitive and rapid in action, such instruments were sometimes deficient in stability. The modern microbalance shown in Figure 40 is both stable and sensitive. It has a 5-inch nickel chromium alloy beam that is light, rigid, and requires neither plating nor lacquering. The beam, end planes and stirrups, jewelled double action arrestment, and internal weight system are enclosed in a separate compartment to minimise the effects of air currents when the lower weighing compartment is opened. Each division of the optical scale is equivalent to two micrograms.

Assay balances, designed to weigh beads of gold or other precious metals, are often of quite light construction. Some forms are portable and can therefore be used for work in the field. The balance depicted in Figure 41 is of this type, and is provided with a rider and a set of weights ranging from one to one thousand milligrams. The balance, which is sensitive to about one-hundredth of a milligram, is shown ready to be transported as soon as the front of the case has been closed. A brass clip held by a thumbscrew prevents the beam from moving, while the pans and their wires have been removed from the beam and placed in separate turnbuckle clips. Some interesting light-load microbalances were in use before the general availability of commercial instruments. One dating from about 1910 is shown in Figure 42, and was used by Whytlaw-Gray and Ramsay to determine the atomic weight of radium[41]. The beam, made by fusing together thin silica rods, turns on a silica knife A that rests on a plate B of polished quartz. One end of the beam carries a fine silica fibre terminating in a hook. This supports the silica weights, sealed quartz buoyancy bulb E of known volume, and a little pan or bucket G. These items are counterbalanced by a single piece of quartz suspended from the other end D of the beam, so that the balance operates on the constant-load principle. A tiny mirror on the beam allows swinging to be observed as the movement of a spot of light on a scale that is outside the vacuum-tight case.

When the object to be weighed has been placed on the pan, the silica weights are adjusted until an approximate state of equilibrium has been obtained. The case is then closed and the pressure within it is adjusted to obtain exact balance. Suppose that the case has been completely evacuated and that the arm carrying the object is now slightly tilted downwards. If some air is let into the case, the bulb (which may be evacuated or may contain air) is buoyed up a little, thus tending to restore the

40 Modern microbalance
(L. Oertling, Ltd.)

beam to the horizontal position. If the temperature is kept constant, the result can be calculated from pressure measurements.

Another type of microbalance was also developed in the early years of the present century. There are no knives and planes, and the restoring force is torsional rather than gravitational[42]. The massive appearance of a modern form of this type of microbalance (Figure 43) is mainly due to the metal casing needed to ensure uniformity of temperature of the beam and torsion elements. Although the capacity is only 250 milligrams in each pan, a standard deviation of better than 0.08 microgram can normally be achieved. Three numbered dials, an engraved drum, and an optical scale allow direct readout of the result. The beam A (Figure 44) is made of thin quartz rods, weighs about 50 milligrams, and is about 4 inches long. Most of the load is supported by two vertical quartz fibres BB, which are only 4 microns (0.004 millimetre) in diameter. In an actual weighing, the object is counterpoised to within one milligram. Spindle C, which drives the readout dials and is also attached to one end of the quartz torsion fibre D, is then rotated until the twisting of this fibre restores the displacement of the beam to zero.

Despite its high sensitivity, the modern torsion microbalance has considerable tolerance to vibration and can, in fact, be used in routine work. Such an instrument

41 Portable assay balance

is almost indispensable in the rapidly-advancing field of submilligram analysis, where the sample may weigh less than one hundred micrograms.

Another way of balancing the load on a microbalance is by the torque produced when a current flows in a coil that is suitably placed in a magnetic field. The movement of an ordinary milliammeter was used in a microbalance designed by the British workers Hales and Turner in 1956[33]. A tiny pan is suspended from a beam of fine aluminium tubing that replaces the meter needle, and the beam is then adjusted to lie horizontally. The placing of a small object upon the pan causes this to sink from its normal position. A current is passed through the meter coil and is adjusted until the beam is again horizontal. Since the torque that is acting as the restoring force is proportional to the current, the act of weighing involves the measurement of the current required to bring the beam back to its original position.

Commercial instruments based on this general principle are now available. Three examples are shown in Figures 45, 46 and 47. The working principles of the RH Electrobalance manufactured in the United States by the Cahn Instrument Company are indicated in Figure 49. An object, such as a small dish, that is placed on the pan marked SAMPLE may be counterbalanced or tared approximately by weights placed on the other pan. The exact balance is achieved by the operation of the photocell. Any deviation of the beam from a pre-determined position causes a change in the amount of light falling on the photocell from the light source. A servo mechanism then operates which causes current to flow in the coil to restore the balance. The current change is followed by a chart recorder which is calibrated appropriately. This balance can weigh amounts of from a few micrograms up to one hundred grams, and can detect weight changes of from one tenth of a microgram up to twenty grams.

In the microbalance manufactured by C. I. Electronics Ltd., of Salisbury, Wiltshire, the weight of the sample is indicated by a meter on the front of the case. This instrument has a range switch that allows amounts up to one hundred milligrams to

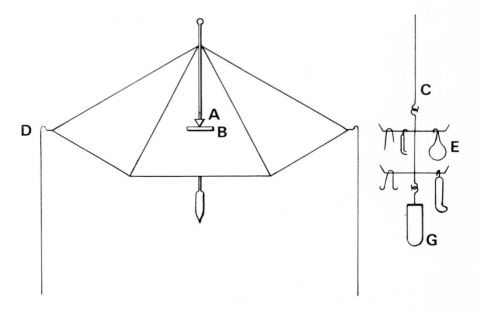

42 Features of the microbalance used by Whytlaw-Gray and Ramsay

be weighed. Balances of this type can be constructed as a single unit, or the 'weighing head' may be at some distance from the control and readout equipment, as in the case of the Cahn RH balance. The small weighing head is easily enclosed in a glass or similar chamber, so that operations can be carried out in a vacuum or in an atmosphere other than that of air.

43 Quartz-fibre microbalance
(L. Oertling, Ltd.)

44 Diagrammatic view of interior of the Oertling quartz-fibre microbalance

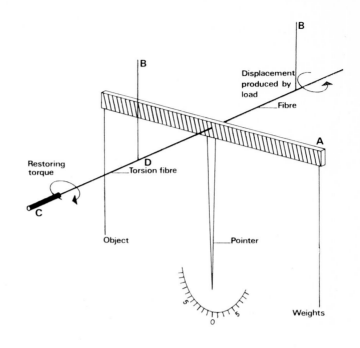

45 Microbalance
(C. I. Electronics Ltd.)

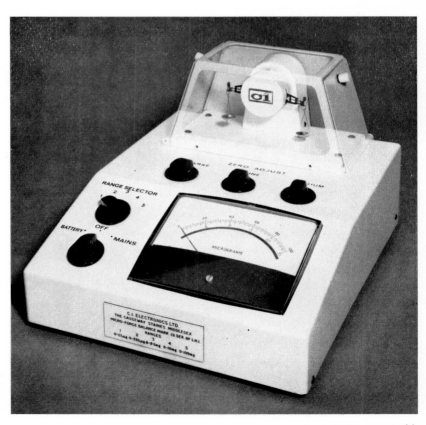

46 Cahn Electrobalance
(Cahn Instrument Co.)

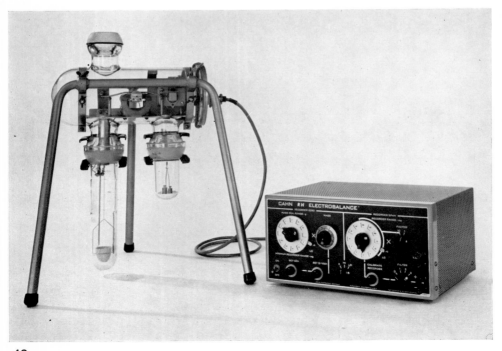

47 Cahn RH Electro-
balance and control
unit
(Cahn Instrument Co.)

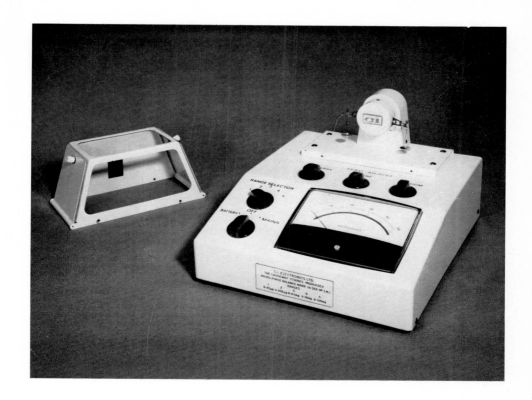

48 Weighing head of Micro-balance
 (C. I. Electronics Ltd.)

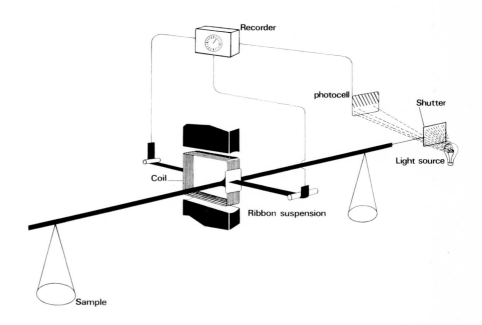

49 Working arrangements of Cahn RH Electro-balance

8 Recording Balances

It is sometimes useful to determine changes in weight undergone by a substance over a period of time. For example, if a technical material such as a propellent explosive or a sample of coal decomposes by giving off a gas which can escape, the decomposition can be studied by following the loss of weight under suitable temperature and other conditions.

One method of determining calcium in, for example, chalk involves dissolving the weighed sample to make a faintly acid solution, precipitating or throwing down the calcium as calcium oxalate, then collecting, washing, and heating the precipitate. Finally, the cooled precipitate is weighed and the amount of calcium is calculated from the result. The conduct of the heating stage is quite important, since some water is retained if the temperature is too low, while partial or complete decomposition into calcium carbonate, or even into calcium oxide, occurs on strong heating. The determination of an element or other constituent by precipitation as a suitable compound which is then weighed is a process known as *gravimetric analysis*. Many such analyses have been in use for a century or more and are described in standard works on analytical chemistry. Until quite recently, many of these descriptions were curiously unbalanced. Although certain steps, such as that of actual precipitate formation, were meticulously specified, the heating of the precipitate to constant weight was sometimes described only in loose phrases, such as 'heat to dull redness for one hour.' Beginning about 1950, the French analytical chemist Clément Duval has systematically examined the effect of heating on hundreds of analytical precipitates, and has been able to specify closely the conditions that give the most satisfactory results[44]. Thermal methods are now important analytical techniques[45].

Duval improved the *thermobalance,* an instrument developed in 1944 by Chevenard and his co-workers[46]. Essentially, the device is an aperiodic balance with a silica rod projecting upwards from one end of the beam. The vessel, such as a crucible, that contains the precipitate is mounted on the upper end of the rod, which projects into an electric furnace without touching the wall or bottom. The other end of the beam carries a small mirror. This causes a spot of light to fall upon photographic paper that is mounted on a clock-driven vertical drum. The clock also operates a device that causes the temperature of the furnace to rise steadily over a period of up to several hours. Normally, the precipitate first loses weight quite rapidly as moisture is driven off, so that the spot of light traces a steep line on the paper. If conditions of stability to heat are then reached, the trace becomes horizontal. If a temperature is reached at which further loss in weight occurs, the trace slopes once more until the residue in the crucible has a composition that is stable at the now higher temperature. An inspection of the trace enables suitable heating rates, times, and temperature limits to be suggested for actual analytical use.

Electronically controlled recording balances date from 1938, when the American

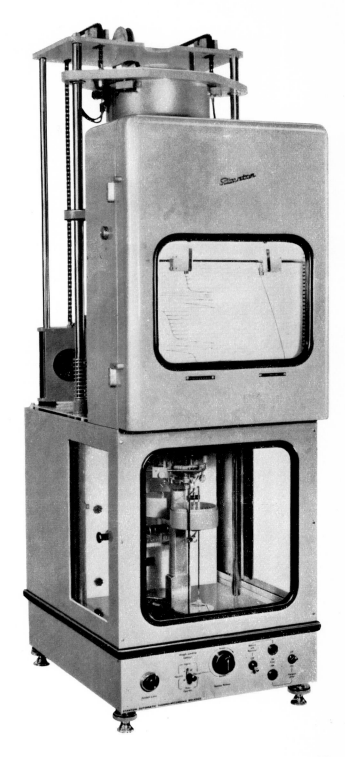

50 Thermo-recording balance
(Stanton Instruments, Ltd.)

workers Muller and Garman described an instrument that is based upon a 'chainomatic' balance[47]. The pointer carries a shutter that interrupts the illumination of a photoelectric cell. This cell actuates an electric motor, which restores the state of balance by appropriately shifting the chain carriage. The range over which automatic operation is obtained is of course limited to the change in weight covered by the full extent of travel of the carriage.

Figure 50 shows a modern thermo-recording balance made by Stanton Instruments, Ltd. The design is based upon an aperiodic balance of the constant-load type, with a vertical silica or alumina sample-carrying rod projecting upwards into the furnace. When the sample weight begins to change, a plate mounted on the beam moves slightly. The gap, and hence the electrical capacity, between this plate and another that is mounted on a motor-driven arm is thereby altered. The electrical unbalance thus created automatically energises the motor. This restores the original gap between the plates and, at the same time, drives the recording pen to indicate the change in weight. The trace of a second pen records the change of furnace temperature with time.

This instrument can accommodate weight changes several times greater than that required to produce maximum beam deflection or full-scale travel of the weight-indicating pen. Instead of going off-scale, a deflection that is reaching the maximum is returned to zero by the automatic loading of a small weight. Repetition of this operation enables a total weight change of up to one gram to be automatically followed and recorded on a chart width corresponding to only one hundred milligrams.

Although the most obvious use of a vacuum balance is to eliminate small errors otherwise caused by the buoyancy effect of air (see page 28), the loss of moisture or of other volatile substances from a material such as paper can be followed by continuous weighing in a vacuum. Observations of this kind are now quite important, since the sample is under conditions that simulate those of outer space. A vacuum recording balance is particularly useful for such studies. In the Ainsworth instrument shown in Figure 51, pumps, valves, vacuum gauging gear, and recording equipment take up much more room than the balance proper, which is within the bell-shaped glass housing at the right.

51 Recording high-
vacuum balance
(Wm. Ainsworth & Sons, Inc.)

References

1 Norton, Thomas (fl. 1477), 'Ordinall of Alchimy,' British Museum, Addit. MS 10302, f.37.

2 Black, J., 'Experiments upon magnesia alba, quicklime and some other alcaline substances,' *Essays and observations, Physical and Literary, Read before a Society in Edinburgh and Published by them,* 1756, Vol. 2, p. 157; Alembic Club Reprints, No. 1, Edinburgh, 1893 onwards.

3 Freund, I., 'The Study of Chemical Composition,' Cambridge University Press, 1904, p. 62.

4 Landolt, H. H., 'Über die Erhaltung der Masse bei chemischen Umsetzungen,' *Abhandlungen der Akademie der Wissenschaften zu Berlin, Physikalisch-mathematische Klasse,* Abh. 1, Berlin, 1910.

5 Partington, J. R., 'General and Inorganic Chemistry,' MacMillan, London, 1946, p. 252. See also Moore, W. J., *Journal of Chemical Education,* 1961, Vol. 38, p. 232.

6 Cameron, A. E., 'The determination of atomic weights by mass spectrometry,' *Analytical Chemistry,* 1963, Vol. 35, No. 2, p. 23A.

7 Wichers, E., 'How good are the new atomic weights?' *Analytical Chemistry,* 1963, Vol. 35, No. 3, p. 23A.

8 Walker, J., 'The Theory and Use of a Physical Balance,' Clarendon Press, Oxford, 1887, p. 10.

9 Dittmar, W., in 'Thorpe's Dictionary of Applied Chemistry,' ed. Thorpe, J. F. T., and Whiteley, M.A., Longmans, Green & Co., London, 1937, Vol. 1, p. 592.

10 MacNevin, W. M., 'The Analytical Balance: its Care and Use,' Handbook Publishers, Inc., Sandusky, Ohio, 1951, p. 14.

11 Hauksbee, F., in 'Lexicon Technicum, or an Universal English Dictionary of Arts and Sciences,' by Harris, J., London, 1710, at p. Hhh.

12 Kater, H., 'A letter respecting the construction of a balance,' *Quarterly Journal of Science, Literature, and the Arts,* 1822, Vol. 12, p. 40; see also the same journal, 1821, Vol. 11, p. 280.

13 Thorpe, E. (Editor), 'The Scientific Papers of the Honourable Henry Cavendish, F.R.S.,' Cambridge University Press, 1921, Vol. 2, p. 56, footnote.

14 Dittmar, W., in 'Thorpe's Dictionary of Applied Chemistry,' ed. Thorpe, J. F. T., and Whiteley, M. A., Longmans, Green & Co., London, 1937, Vol. 1, p. 587.

15 Barclay, A., 'Pure Chemistry,' Science Museum Handbook, H.M. Stationery Office, London, 1937. Part I, Historical Review, p. 29; Part II, Descriptive Catalogue, p. 56.

16 Barclay, A., 'Some early chemical balances,' *Journal of the Society of Chemical Industry,* 1935, Vol. 54, supplement to issue of October 18, p. S2.

17 Daumas, M., 'Les Instruments Scientifiques aux XVIIe et XVIIIe Siècles,' Presses Universitaires de France, Paris, 1953, p. 292.

18 Cavendish, H., 'Experiments on air,' *Philosophical Transactions,* 1784, Vol. 74, p. 119; Alembic Club Reprints, No. 3, Edinburgh, 1893 onwards.

19 Daumas, M., 'Lavoisier, Théoretician et Expérimentateur,' Presses Universitaires de France, Paris, 1955, p. 134.

20 Lee, S. (Editor), 'Dictionary of National Biography,' Smith, Elder & Co., London, 1896, Vol. 47, p. 265.

21 Blagden, C., 'Report on the best method of proportioning the excise on spirituous liquors,' *Philosophical Transactions,* 1790, Vol. 80, p. 321.

22 Lee, S. (Editor), 'Dictionary of National Biography,' Smith, Elder & Co., London, 1899, Vol. 57, p. 259.

23 Taylor, E. W., Wilson, J. S., and Maxwell, P. D. S., 'At the Sign of the Orrery,' Vickers Instruments, York (undated), p. 25.
24 Shuckburgh Evelyn, G., 'Endeavours to ascertain a standard of weight and measure,' *Philosophical Transactions,* 1798, Vol. 88, p. 139.
25 Lee, S. (Editor), 'Dictionary of National Biography,' Smith, Elder & Co., London, 1898, Vol. 53, p. 272.
26 Daumas, M., 'Les Instruments Scientifiques aux XVIIe et XVIIIe Siècles,' Presses Universitaires de France, Paris, 1953, p. 336.
27 Gunther, R. T., 'Early Science in Oxford,' Oxford, 1923, Vol. 1, p. 76.
28 Child, E., 'The Tools of the Chemist,' Reinhold, New York, 1940, p. 81.
29 Kater, H., 'An account of the construction and adjustment of the new standards of weights and measures of the United Kingdom of Great Britain and Ireland,' *Philosophical Transactions,* 1826, Vol. 116, p. 1.
30 Schumacher, H. C., 'A comparison of the late imperial standard troy pound weight with a platina copy of the same, and with other standards of authority,' *Philosophical Transactions,* 1836, Vol. 126, p. 457.
31 Miller, W. H., 'On the construction of the new imperial standard pound and its copies of platinum, and on the comparison of the imperial standard pound with the Kilogramme des Archives,' *Philosophical Transactions,* 1856, Vol. 146, p. 753.
32 Kater, H., and Lardner, D., 'Mechanics,' The Cabinet Cyclopaedia, London, 1830, p. 286.
33 Dittmar, W., in 'Thorpe's Dictionary of Applied Chemistry,' ed. Thorpe, J. F. T., and Whiteley, M.A., Longmans, Green & Co., London, 1937, Vol. 1, p. 588.
34 Oesper, R. E., 'Some famous chemical balances,' *Journal of Chemical Education,* 1940, Vol. 17, p. 312.
35 'Reports of the Juries, Exhibition of the Works of Industry of All Nations, 1851,' Spicer, London, 1852, Vol. 2, p. 556.
36 Bunge, P., 'Neue Construktion der Wage,' *Repertorium für Physikalische Technik, für Mathematische und Astronomische Instrumentenkunde,* 1867, Vol. 3, p. 269.
37 Dunbar, M., 'Some modern chemical balances,' *Journal of the Society of Chemical Industry,* 1935, Vol. 54, supplement to issue of October 18, p. S10.
38 Raudnitz, M. (revised Reimpell, J.), 'Handbuch des Waagenbaues: Band I, Handbediente Waagen,' Verlag B. H. Voigt, Berlin, 1955.
39 Felgentraeger, W., 'Feine Waagen Wägungen und Gewichte,' Springer, Berlin, 1932.
40 Ingram, G., 'Methods of Organic Elemental Microanalysis,' Chapman & Hall, London, 1962.
41 Whytlaw-Gray, R., and Ramsay, W., 'The Atomic weight of radium,' *Proceedings of the Royal Society,* 1912, Section A, Vol. 86, p. 278.
42 Manley, J. J., in 'Thorpe's Dictionary of Applied Chemistry,' ed. Thorpe, J. F. T., and Whiteley, M. A., Longmans, Green & Co., London, 1937, Vol. 1, p. 607.
43 Hales, J. L., and Turner, A. R., 'An inexpensive microbalance,' *Laboratory Practice,* 1956, Vol. 5, p. 245.
44 Duval, C., 'Inorganic Thermogravimetric Analysis,' Elsevier Publishing Co., Amsterdam and London, 2nd edition, 1963.
45 Wendlandt, W. W., 'Thermal Methods of Analysis,' Interscience Publishers (John Wiley & Sons), New York, 1964.
46 Duval, C., 'Continuous weighing in analytical chemistry,' *Analytical Chemistry,* 1951, Vol. 23, p. 1271.
47 Muller, R. H., and Garman, R. L., 'An electronic recording analytical balance,' *Industrial and Engineering Chemistry, Analytical Edition,* 1938, Vol. 10, p. 436.

Appendix

Inventory Numbers of Science Museum objects shown in this survey

Figure Number	Inventory Number
2	1915–34
4	1914–858
7	1936–14
8	1961–24
10	1930–770
14	1900–166
17	1876–227
19	1926–810
25	1957–11
29	1932–580
30	1876–369
31	1876–380
34	1935–284
35	1959–310
38	1964–40
41	1955–173
42	1947–265
43	1964–464
45	1966–177
46	1966–273
47	1966–274